Evgeny Shnyukov

Black Sea mud volcanoes as a search criterion for gas hydrates

Evgeny Shnyukov

Black Sea mud volcanoes as a search criterion for gas hydrates

ScienciaScripts

Imprint
Any brand names and product names mentioned in this book are subject to trademark, brand or patent protection and are trademarks or registered trademarks of their respective holders. The use of brand names, product names, common names, trade names, product descriptions etc. even without a particular marking in this work is in no way to be construed to mean that such names may be regarded as unrestricted in respect of trademark and brand protection legislation and could thus be used by anyone.

Cover image: www.ingimage.com

This book is a translation from the original published under ISBN 978-613-4-91313-3.

Publisher:
Sciencia Scripts
is a trademark of
Dodo Books Indian Ocean Ltd. and OmniScriptum S.R.L publishing group

120 High Road, East Finchley, London, N2 9ED, United Kingdom
Str. Armeneasca 28/1, office 1, Chisinau MD-2012, Republic of Moldova, Europe
Printed at: see last page
ISBN: 978-620-8-09766-0

TABLE OF CONTENTS

Introduction

Mankind is on the threshold of development of a new and practically inexhaustible type of energy raw material - methane hydrates. Nowadays, the grandiose scales of methane gas hydrates distribution in the subsurface have already been established, estimated by figures from IO^{18} m^3 to IO^{19} m^3 . Methane resources in gas hydrates are almost 2 orders of magnitude higher than conventional methane resources. However, the development of methane in gas hydrates is a complex and difficult task. Countries with limited hydrocarbon resources - Japan, China, India, USA, South Korea, etc. - are leading in this work. In recent years, the EU countries have become interested in this subject. Many of these countries have created their national programs for the development of methane gas hydrates and allocated significant funds for this purpose. It was expected that Japan would start gas hydrate methane production in 2OO7. These dates were postponed to 2O17 and then to 2O25. Unexpectedly, in May 2O17, China announced the successful production of methane from gas hydrates, thus gaining leadership in the study and development of methane gas hydrates. I would like to emphasize that the development of gas hydrates is a technically challenging task. But before we start to develop gas hydrates, they still need to be found. It is generally accepted in the literature that 98% of natural gas hydrates are located within the World Ocean. There is literature data that about 1O% of the World Ocean area is hydrate-bearing.

Geological and geophysical study of methane gas hydrates is rather expensive and complicated. In this regard, the prospecting criteria that allow narrowing the areas of work are important. Most often they are confined to the continental slope, fault zones, and river cones. In our opinion, an important place of gas hydrates occurrence are mud volcanoes.

Gas hydrate accumulations are formed directly in sediments, in conditions thermodynamically favorable for their occurrence (temperature, pressure). Such conditions exist in different regions of the World Ocean on the shelf, on the continental margin, in deep water zones. The task of their search is the most important task of marine geology today. In this respect, the study of mud volcanoes is one of the ways to study gas hydrates. In the Black Sea, methane gas hydrates have been found in 60% of mud volcanoes located in the thermodynamically favorable zone for hydrate formation - deeper than 600 meters. It is very likely that they are developed in deep sea mud volcanoes everywhere and their detection is a matter of time.

The main provisions of the work are based on the results of geological studies of the Azov-Black Sea basin, and in particular mud volcanoes, conducted during 1970-2015. The research was based on the results of studies of mud volcanoes in the Kerch-Taman region in 19601985 (E.F.Shnyukov, P.I.Naumenko, Y.S.Lebedev et al., 1971; E.F.Shnyukov et al., 1986, 2005) and on the results of marine expeditions to the Black Sea. The work was carried out on the research vessels "Mikhail Lomonosov", "Akademik Vernadsky", "Ichtiander", "Kiev", "Vladimir Parshin", "Geochemist" in the period from 1986 to 2013. Dozens of voyages were made, materials from many hundreds of gravity tubes and dredges were sampled and studied by methods of lithology and mineralogy. The bottom topography, the relief of mud volcanoes, the nature of their emissions, and the composition of gases were studied. Dozens of new mud volcanoes were discovered and studied. In general, the volume of the performed research is very significant.

Mud volcanoes of the Black Sea

Representations of mud volcanoes have changed along with expanding data on their structure.

The first literary definitions reflected only their morphological features. Thus, the "Geological Dictionary" (M., 1973) provides the following definition:

"Mud volcano is a large hill of flat-conical shape, composed entirely or only from the surface of the soprano deposits and possessing at the top a funnel-shaped crater and a channel going to the depth, from which periodically or continuously emit gas, water, sometimes and soprano *mud*. Episodic violent changes are noted."

This definition seems to us fair, but not complete, not reflecting the nature of volcanism.

More profound is the formulation of I.M.Gubkin (1937): "gas-oil manifestations and mud volcanism are a function of the same causes, namely functions of geologic structure, in particular special forms of tectonics - diapiric structures". V.G.Bondarchuk (1959) expressed this idea more succinctly: "mud volcanism is one of the manifestations of tectonics mainly of diapir type".

S.F.Fedorov (1939) considered the presence of diapiric structures as an obligatory condition, even a law for the formation of mud volcanoes in the Crimean-Caucasian region. P.P.Avdusin (1948) actually supported the definitions of I.M.Gubkin.

M.K.Kalinko (1964) noted the existence of the connection of mud volcanoes with diapirs and cryptodiapirs, but recognized it as private. In his opinion, volcanoes arise under the influence of hydrodynamic conditions characteristic of certain tectonic zones. At the same time, as early as in the 30s of the last century, mud volcanoes confined to large disturbances were noted in Azerbaijan (A.A.Yakubov, B.V.Grigoryants, Ad.A.Aliyev et al. 1980).

Mud volcanoes of the Black Sea are diverse deep structures. Significant seismic materials obtained in recent years during oil and gas exploration allow us to fix the roots of these morphostructures in the Mesozoic, basement and even in the upper mantle, below the Moho surface.

In the light of new seismic materials in the Black and Caspian Seas, the question of the deep emplacement of mud volcanoes is quite rightly raised by D.F.Ismagilov et al. (2006) about the deep emplacement of mud volcanoes: "how can clay Maikop diapirs influence the occurrence of mud volcanoes with roots in Mesozoic sediments?". These authors note: "Mud volcanoes are more complex geologic phenomena than commonly thought, not related to the formation of clay diapirism."

In our opinion, a significant part of the mud volcanoes of the Black Sea are still connected with clay diapirs, but this connection is opposite to the generally accepted one. The key to understanding this important relationship is pointed out by L.B. Meisner and D.A. Tugolesov in their paper on fluidization deformations in the sedimentary cover of the Black Sea (1997). Deep fluids, penetrating through the sedimentary cover, seem to facilitate the folding process. In other words, it is not diapirs that give rise to mud volcanoes, but on the contrary, deep fluid flows that give rise to mud volcanoes create favorable conditions for diapirs. Sometimes the trace of deep fluid flows is a subvertical geological body, sometimes they penetrate to the surface through large discontinuities in the continuity of the sedimentary cover and create mud volcanoes.

The obvious participation of deep processes in the formation of mud volcanoes raises the question of the need to clarify the definition of this phenomenon. Probably, the definition of S.A.Kovalevsky (1935) is more successful: "mud volcanism is only a particular form of manifestation on the day surface of gas jets flowing from the depth of the Earth's crust".

In a somewhat more modern form, this definition will sound approximately as follows: mud volcanism is a form of manifestation on the surface of the earth of local gas flows of deep hydrocarbons, accompanied by ejections of sopochnaya breccia, debris, water and forming a peculiar form of relief - cone-shaped hills with craters on the tops. Sometimes these are lakes of rounded shape.

In marine conditions mud volcanoes can be observed only when they create positive forms of underwater relief.

Modern marine acoustic surveys provide images of submarine mud volcanoes that are no less visual than surface photographs. On seismic profiles, all volcanoes have a characteristic image and are well distinguished in the seafloor topography.

In section, a mud volcanic edifice usually has the form of a gentle, often truncated cone. The cone is composed of sopochnaya breccia, the flows of which may have several generations. In plan, the described structures are more or less isometric, rounded in shape. The transverse dimensions of mud volcanoes (diameter of their base) vary widely, from the first hundred meters to several kilometers. Thus, in the central part of the Black Sea, the transverse size of mud volcanoes reaches 1.5-2.5 km and even 4 km. The height of mud volcanoes is relatively small in comparison with their diameter: even the largest of them rarely exceeds 100 meters. Therefore, the steepness of the slopes of mud volcanoes, as a rule, does not exceed the first degrees. This ratio of diameter to height is explained by the low density of the sopochnaya breccia, which can spread over a distance of several kilometers.

Mud volcanoes are found in many of the major geological structures of the Black Sea. They are abundant in the Sorokina Trough, the West Black Sea Basin, the Kerch-Taman Trough and others. They usually have a well-designed crater, with a cross-section of several tens to hundreds of meters. In addition to the main crater, small secondary outcrops of liquid and gas components, called salsas and gryphons, are often found on the slopes of mud volcanoes.

It is necessary to distinguish mud volcanoes from gas flares, which are numerous on the seafloor in some regions. Mud volcanoes are particular morphostructures of seafloor sediments, often having geomorphologically distinct positive seafloor forms. Gas flares are less significant releases of gases, usually confined to rock discontinuities and not forming positive shapes on the seafloor.

The gases of the Black Sea mud volcanoes are 95-99% methane. Ethane, propane, butane, pentane and hexane are also present. Impurities of hydrogen sulfide, CO_2 , nitrogen are often recorded. Findings of hydrogen sulfide in mud volcanoes are of interest. So, on the volcano Mitina, during our expeditions Y.N.Goryachkin revealed an anomaly in distribution of hydrogen sulfide concentrations, in particular, absolute increase of hydrogen sulfide content; salinity anomalies and a peculiar rise of isosurfaces of heat transfer in the form of a dome. Sharp odor of hydrogen sulfide was recorded in geological columns at the volcanoes Neftyanoy, Volokhina, Irkutskiy, Platformenny, Udodova, Trakia and others. The volcanic sediments often contain thick interlayers enriched with hydrotroilite. All this allows us to assume the participation of mud volcano outbursts in the formation of hydrogen sulfide contamination of the Black Sea.

The activity of mud volcanoes is accompanied by outbursts of huge methane masses. Thus, several mud volcanoes were encountered in the Sorokin depression, with gas fountains at which the height was 800-1300 m (Dvurechensky, Vodyanitsky, Kazakov, Helgoland, etc.). The fountain at Dvurechensky volcano had a diameter of 400 m (E.F. Shnyukov et al., 2013) (Figs. 1, 2). Given the rarity of volcano observations in the sea, it is necessary to compare marine mud volcanoes with terrestrial volcanoes to fully understand the phenomena. In Azerbaijan, for example, according to R.R. Rakhmanov (1987), an average of 250 million m^3 of methane is emitted per volcanic eruption, and 125×10 м ![123]

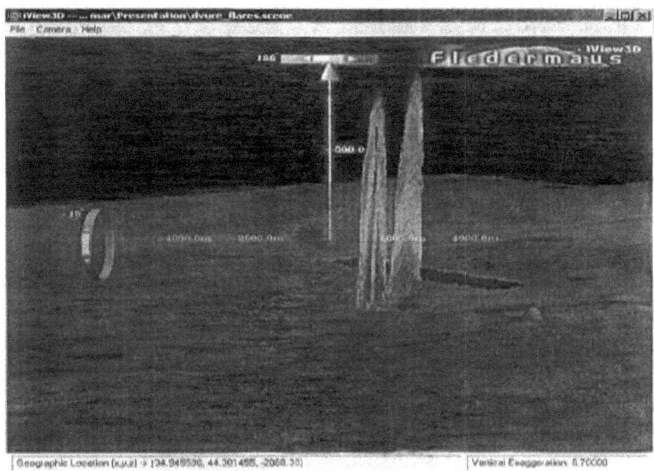

Fig. 1. Three-dimensional image of gas emissions in the area of the Dvurechensky mud volcano (by

Yu.G. Artemov, with the participation of Dr. Ens Greinert (Geomar, Germany). Bottom measurement data were obtained during cruise M 52/1 of the R/V Meteor, echogram of fountains - during the 57th cruise of the R/V Meteor

"Professor Vodyanitsky" (2002).

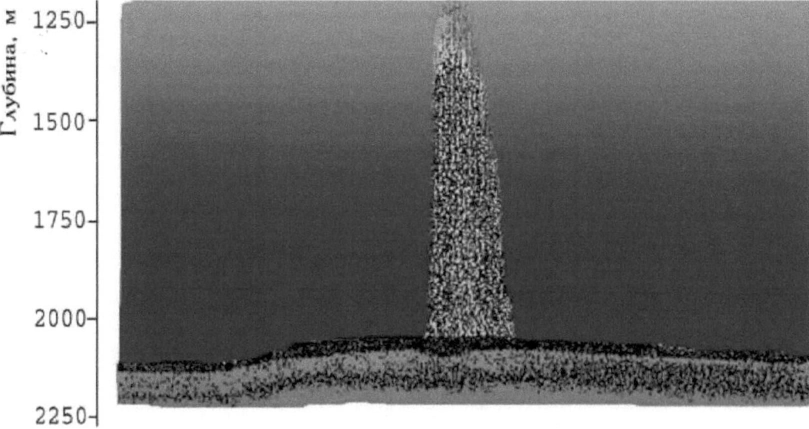

Fig. 2. Echogram of a gas fountain with a height of about 800 m (above the seafloor) at the Dvurechensky mud volcano (2003). Materials of the 59th cruise of the R/V "Professor Vodyanitsky".

The overall scale of this process has been assessed in detail by Ad.A.Aliyev et al. (2015) especially on the example of Azerbaijan.

A.M.Plotnikov (1967) notes that the single-act gas emissions of the Kerch volcanoes Dzhau-Tepe and Dzhardzhava **were** 88×10^6 and 6.2×10^6 m^3 respectively per one eruption, etc. (Figs. 3, 4, 5).

Fig. 3. Explosive outburst of the mud volcano Dzhau-Tepe (E.Steber, 1909).

Fig. 4. The 1982 eruption of the mud volcano Dzhardzhava (picture by artist V.I.Kantemirov).

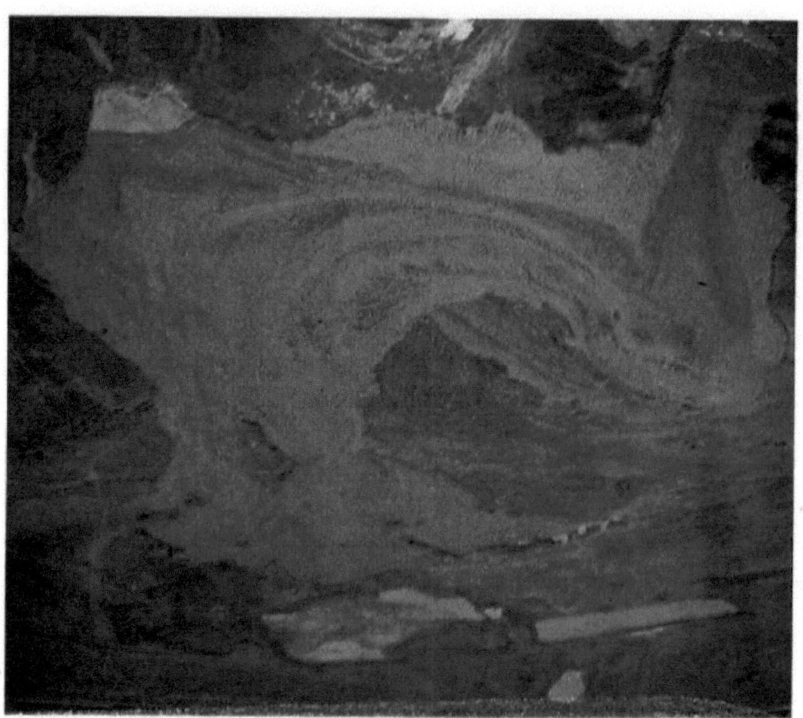

Figure 5. Mud flow from the eruption of Jardzhava volcano, 1982.

The magnitude of gas emissions from Black Sea volcanoes has not yet been estimated, but they are certainly commensurate with those of their land-based counterparts. Mud volcanoes do not only spew large volumes of gases to the surface: they eject huge masses of sap breccia and clastic material from the erupted rocks. According to G.A.Lychagin (1952), the Kerch mud volcanoes ejected at least 25 billion m^3 of sap breccia during their geological activity.

P.I.Naumenko (1976) increases this figure up to 40 billion m^3 , taking into account the materials of long-term drilling works on the Kerch volcanoes.

Only one mud volcano Dzhau-Tepe erupted - 55 million m^3 breccia, etc.

Examples of this kind can be continued, but it is probably already clear that the volume of ejected masses is enormous.

H. A.Golovkinskii (1889), S.E.Alyaev (1947), Z.L.Maimin (1951), G.A.Lychagin

(1952), E.F.Shnyukov and others. (1971) came to the conclusion that as a result of all these processes on the tops of anticlinal structures affected by mud volcanism, weakened areas - a kind of mass deficit, where hydrostatic pressure is distributed unevenly - appear. The ejected sap breccia presses on the weakened area, which leads to large subsidence. The example of the mud volcano Lokbatan (Gavrilov, 1939) is quite illustrative in this respect. Gas emission, which continued here for several years, apparently caused compaction of the sopochnaya breccia and led in 1933 and 1935 to the formation of a series of huge sinkholes and subsidence failures, each subsequent one overlapping the previous ones. The last sinkhole was of grandiose size.

Similar phenomena were described after the eruptions of the mud volcano Dzhau-Tepe in 1909 and 1914. Under the conditions of submarine eruption, the accumulation of sediments in a cavity or sinkhole would increase the process of subsidence, resulting in the continued growth of deflections and an increase in the volume of sediments accumulated in them. Subsidence accompanies all mud-volcanic eruptions without exception. As a matter of fact, the subsidence of the Earth's surface as a result of the removal of matter from the interior is not an exclusive property of mud volcanoes alone. The same is characteristic of volcanoes proper. Back in the late XIX century, the Austrian volcanologist Reyer came to the conclusion that eruptions and removal of lava to the surface occur due to the lowering of the Earth's crust (Kovalevsky, 1935). Collapses of volcanoes after eruptions are widely known and described in many geological works.

According to K.A. Prokopov (1931), after formation, an anticlinal fold becomes an arena of intense mud volcanism and in certain areas begins to sag as a result of the ejection of significant masses of mud to the surface. The fold develops very slowly and smoothly, which determines the plastic bending of the underlying strata and its bowl-shaped structure. K.A. Prokopov (1931) shows the stages of development of compensatory synclines in the form of a series of schemes (Fig. 6). The initial structure is represented as a stratum of plastic clay breccia-like mass, with the formation of basin craters and dragging of subsequent younger sediments into them. The depressions composed by more resistant rocks "resisted the folding forces and deformed in a very complicated way, not withstanding the general pattern of tectonics of the area".

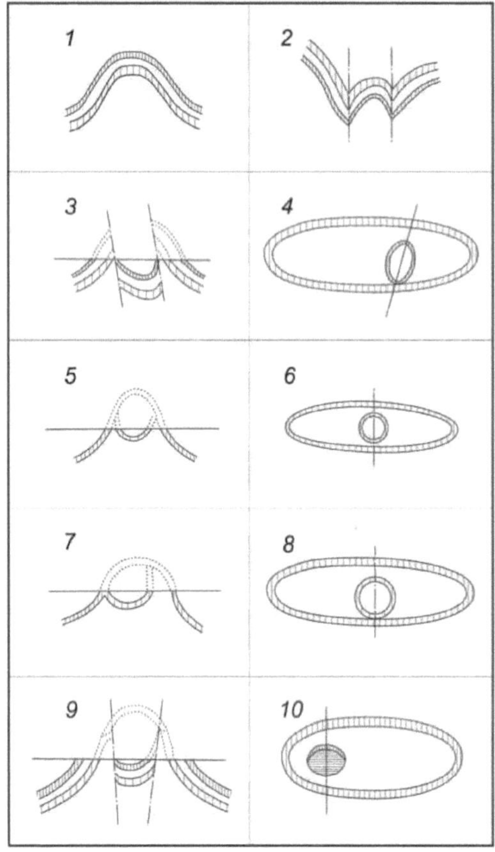

Fig. 6. Stage of development of compensatory synclines.

By K.A.Prokopov (1931)

G.A.Lychagin (1952) clearly formulated the thesis about the origin of the Kerch depressed synclines due to the activity of mud volcanoes in certain paleogeographic conditions.

Almost all modern authors emphasize the mud-volcanic origin of the compensation troughs, synclinal location of the complex of Tertiary sediments and sopopochnaya breccia composing them. In this connection, these structures should be called "compensatory synclines" of mud-volcanic origin. This name would reflect both structural and genetic features of all structures of this type in different areas. The term "depressed syncline", widely used and applied by many researchers of the Kerch Peninsula, is unsuccessful. It does not reflect the peculiarities of the genesis. Similarly, the definition of these structures as complications is not quite successful. However, the term "depressed synclines" has been used for many years and has become familiar to a wide range of geologists. In this connection, it seems to us, it is possible to allow its use as a local, Kerch term, using the definition "compensatory syncline" as a more universal and precise one in the general characterization of structures of this type from different regions.

The formation of compensatory synclines occurs only in certain structures that have been tectonically prepared beforehand.

The influx of deep fluids causes the core of the anticline to be saturated with a large amount of high-pressure gases.

Fluid removal associated with mud volcanic activity or tectonic disturbances leads to shrinkage of the core, accompanied by a significant reduction in its volume. This results in the formation of vaulted grabens and similar structures.

The decisive factor in the formation of compensatory synclines is mud volcanism.

At a certain point in the development of the structure, certain parts of the anticline sag in the arch (Fig. 6, *1)* or in its weakened part due to mud volcanism, forming a bowl-shaped sag (Fig. 6, 2). Further deflection leads to the formation of a ring graben (Fig. 6, 3). The variants of development of compensatory synclines are different. They are considered by K.A.Prokopov on the example of the Kerch Peninsula (Fig. 6, *5-10}.* Thus, the Royal structure, lying symmetrically in the center of the anticline, on the axis of the latter, corresponds to (Fig. 6, 5, 6); for Burulkaisky cauldron and other similar structures, where one of the wings of the anticline is connected to the ditch by flexural kink, are characterized by the general features noted in Figs. 6, 7, 8. Finally, some asymmetry of the annular fault with respect to the flexure explains the appearance of half moons of the Kayala-Sarta, Kara-Sidzheut, Kashik and other depressions (Figs. 6,

9, 10).

In our opinion, K.A. Prokopov's scheme of the development of depressed synclines generally correctly reflects the main stages of the evolution of these structures. Extensive factual materials allow us to detail it somewhat. Completion of anticlinal structure formation and manifestation of mud volcanism creates initial prerequisites for the emergence of a compensatory sag. The mud volcano activity inevitably forms a depression in any paleogeographic conditions, in marine and subaerial environment, and leads to bending of rock strata composing the vault part of the anticline. G.A.Lychagin (1952) expressed, however, the opinion that depressions cannot occur in subaerial conditions. The actual materials show that even in the terrestrial conditions there can occur to a greater or lesser extent pronounced depressions.

The further development of the sag is determined by many circumstances. In the sea basin, during underwater eruptions, the sap breccia erupted by the mud volcano together with the accumulated normal marine sediments creates a heavy pressure, which additionally accelerates subsidence and sinking of the areas of anticlinal structures overlying the mud volcanic centers. On land, the erupted breccia is weathered, largely eroded and dispersed, and sedimentation does not actually occur. Therefore, in the subaerial environment, the process of compensatory deflection formation is slowed down and its scale and amplitude are reduced. This process still has often quite significant scales (Bulganak Basin). It is possible to assume such kind of formation with subsequent burial for some depressions of Southwest Turkmenistan, Azerbaijan.

If the marine regime is preserved, the development of the compensatory syncline continues further: as sediments and breccia accumulate, the strata in the anticline vault sink and their continuity is broken. As the form of sinking is most often a rounded compensatory funnel, the form of disturbance is to a greater or lesser extent a pronounced annular discharge formed by a system of smaller disturbances. A peculiar rounded graben filled with sediments and breccia appears. In some places, the mud volcano erupts directly in the compensatory syncline, sometimes along the disturbances framing the compensatory syncline, and often the center of the volcano is displaced away from the compensatory syncline. Sopoic deposits are established in the section of the vast majority of compensatory synclines, indicating the predominance of the first two ratios of mud volcanoes and compensatory synclines. After the retreat of the sea, the depressions were reprecipitated by weathering processes so that they formed positive relief forms - hills, low cliffs, etc. The different lithological composition of the rocks composing the depressions and their intervening structures has had an effect: dense dolomitized limestones occupy a significant place in the composition of the former, while the intervening structures are composed almost exclusively of Maikop clays. By the way, the depressions, in the section of which clays predominate, are not expressed in the relief.

In fact, the development of depressed synclines was immeasurably more complicated, because many of them are an extremely complex "layer cake" of various sedimentary rocks and saphenous breccia. On the periphery of the structures, reef limestones developed, with further sinking eversion to the center, etc. In terrestrial conditions, further activity of mud volcanism leads to continued subsidence and leveling of the terrain. Finally, with further development of the paleogeographic situation towards deepening of the sea and cessation of mud volcanic activity, burial of structures occurs (Anastasievsko-Troitskaya compensatory syncline of the Taman Peninsula). In the conditions of the Kerch Peninsula, most of the depressed synclines stopped in their development at the third stage, because the marine environment was excluded; when the marine environment was preserved, the depressed synclines were buried or continued to develop.

In the conditions of the deep Black Sea, methane gas hydrates may accumulate in the compensation structures of mud volcanoes.

Mud volcanoes in shallow sea conditions

The activity of mud volcanoes in shallow sea conditions does not differ in the manifestation of mud volcanism on land. A clear example of shallow-water volcanic activity is the Golubitsky mud volcano in the Sea of Azov. Thus, only in the current century, it erupted several times (Fig. 7, 8, 9, 10, video frames from the Internet).

Fig. 7. Eruption of the Golubitsky underwater mud volcano in 2002.

Fig. 8. Eruption of the Golubitsky underwater mud volcano in 2008. A frame from video shooting.

Fig. 9. Eruptions of the Golubitsky underwater mud volcano in 2008. A frame from video shooting.

Fig. 10. Eruptions of the Golubitsky underwater mud volcano in 2015. Frame from the video footage.

The subsidence structures of the Golubitsky volcano have not been studied. Methane gas hydrates in the shallow sea are absent due to thermodynamic reasons. But more interesting situations were observed in the geologic past. 4-5 million years ago, in the so-called Cimmerian (Pliocene stage), the area of mud volcanoes activity on the Kerch Peninsula coincided spatially and temporally with the epoch of sedimentary ore accumulation. Iron-ore sediments were accumulated in many subsiding mud-volcanic structures occupied by the shallow Cimmerian Sea, and iron ores appeared. As a rule, their development areas are not large, but their thicknesses are considerable. Iron-ore sediments were accumulated in mud-volcanic structures, which deepened as the sediments accumulated, often accompanied by the appearance of ring and other disturbances. The Baksinskoye iron ore deposit developed in the mud volcano caldera (Figs. 11, 12), Novoselovskoye iron ore deposit and others are quite illustrative examples. Their reserves are quite significant 40 million tons of ore in Baksinskoye, 150 million tons of ore in Novoselovskoye deposit. On the Kerch Peninsula there are 9 iron ore deposits and occurrences localized in subsidence structures directly in the calderas of mud volcanoes or near them. The ores in them are soyed, they are a special type of Kerch oolitic iron ores. They are often interlayered with saponic breccia, occurring most often in deep grabens. Mud-volcanic structures were the reservoir of sedimentary iron ores with total reserves of 400-450 million tons (E.F. Shnyukov and P.I. Naumenko, 1971; E.F. Shnyukov et al., 1991).

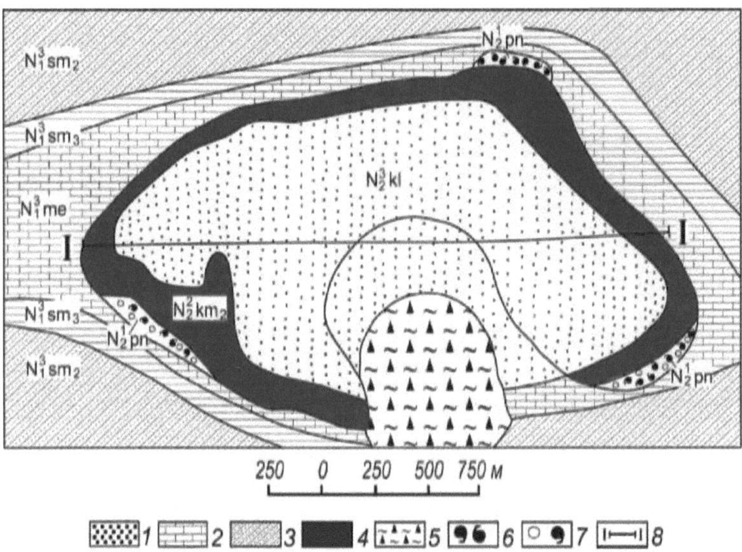

Figure 11. Schematic of the Baksa sopochnoid field and indentation.

According to E.F.Shnyukov, P.I.Naumenko (1964).

Notation: *1* - sands, clayey sands; *2* - limestones, dolomitized limestones; *3* - clays with marl interlayers; 4 - brown iron ore; *5* - sopochnaya breccia; *6* - clayey coquina; 7 - ferruginous coquina; *8* - profile line.

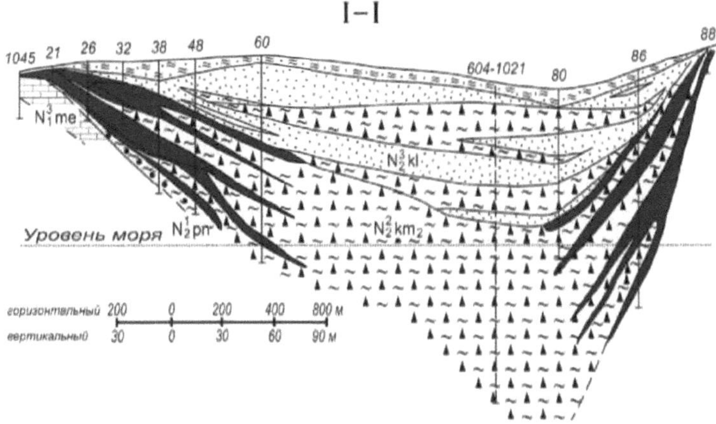

Fig. 12. Geologic section of the Baksinskoye sopochnoe field and indentations along the I-I line. According to E.F.Shnyukov, P.I.Naumenko (1964).

See Fig. 11 *for notation*.

It is not excluded that new finds of ore "indentations" are still possible in the areas of the former shallow sea. The south-east of the Sea of Azov in the area of Kamenny Cape - Kazbek Bank may serve as an example.

A total of 170 mud volcanoes are known in the Black Sea region, including about 100 on land and the rest in the Black and Azov Seas.

Modern mud volcanoes in shallow water conditions have been identified in the Sea of Azov and in the Kerch Strait. The bulk of mud volcanoes in the Azov-Black Sea basin are localized in the deep sea.

In the geological past, mud volcanoes in the shallow sea were quite numerous. These are the territories of the Kerch and Taman peninsulas and obviously part of the Azov Sea coastal waters.

Mud volcanoes of the deep sea

The development of mud volcanoes in **the** deep Black Sea coincides spatially with the zone of methane gas hydrate development (Fig. 13). Due to thermodynamic reasons, the hydrate **formation** zone in the Black Sea is the water area deeper than 600-700 m. In the majority (60%) of the mud volcanoes developed deeper than 600-700 m gas hydrates have been detected as part of the sopochnaya breccia. During many expeditions of various organizations and agencies, methane gas hydrates were repeatedly lifted by bottom tubes. Only a three-meter layer of **sediments** was actually available for study. Gas hydrate neoplasms were found in different intervals of this **three-meter** layer. The column of our stations sampled in the Sorokin depression is given as an example (Fig. 14).

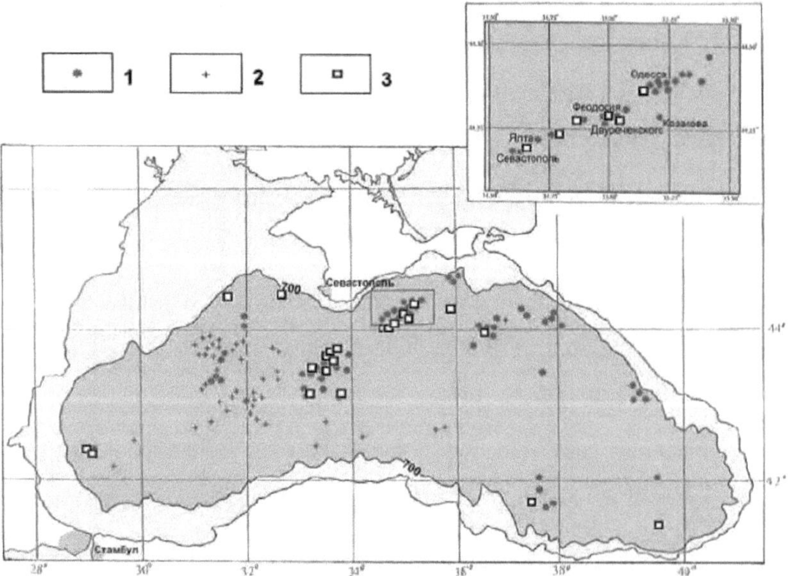

Figure 13: Findings of gas hydrates in the Black Sea.

1 - mud volcanoes; 2 - suspected mud volcanoes; 3 - gas hydrate **finds** in and outside mud volcanoes.

The appearance of methane gas hydrate outcrops is quite diverse. In some mud volcanoes, gas hydrate outcrops have been described in detail. Thus, A.N.Stadnitskaya (2004) considered the morphology of methane gas hydrate outcrops from the sopochnaya breccia of Odessa volcano (Sorokina depression). Gas hydrate aggregates often form irregular tabular fragments up to 4 cm in size, most likely, remnants of veins or interlayers, massive tabular aggregates up to 7 cm in size, a kind of "ice drops" - rounded aggregates no larger than 0.5 cm, surrounding large gas hydrate extractions, etc. (Fig. 15). Sometimes gas hydrate formations are a kind of clumps, as in the Pechori volcano column (Fig. 16) (Borman et al., 2007). Most often, however, gas hydrates are isolated in the form of small crystals of whitish color in the voids of the sopochnaya breccia (Fig. 17).

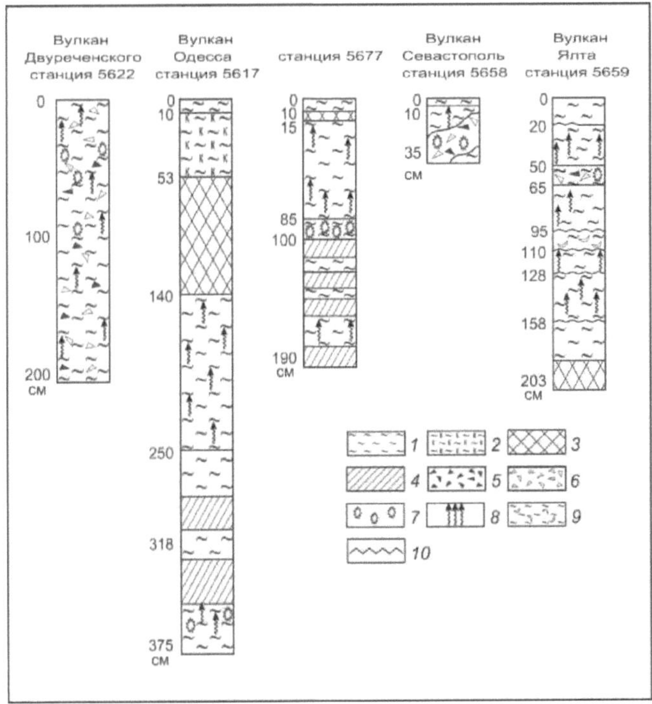

Figure 14: Columns of stations containing gas hydrates.

Notation: *1* - silt; *2* - coccolithic silt; *3* - sapropel; *4* - hydrotroilite silt; *5* - stone breccia; *b* - clay breccia; *7* - gas hydrate inclusions; *8* - degassing; *9* - silt *with* inclusion of fauna; *10* - erosion contact. (Constructed by S.A.Kleshchenko).

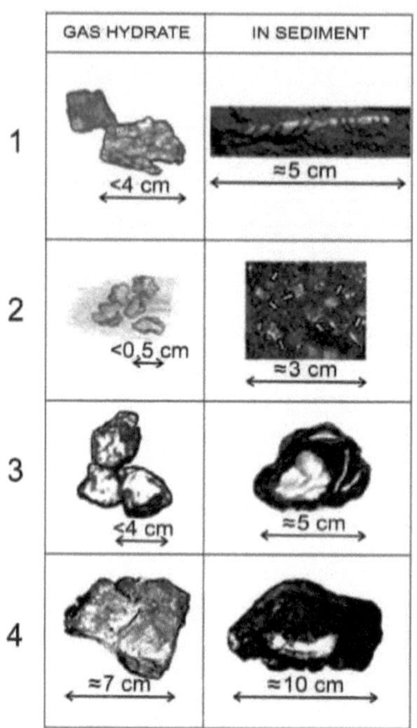

Fig. 15. Morphology of gas hydrate outcrops in the sopochnaya breccia of the Odessa mud volcano. According to A.N. Stadnitskaya (E.F. Shnyukov et al., 2015).

Fig. 16. Character of methane gas hydrate outcrops in the sediment column of the Pechori mud volcano. Light-colored areas of the core are gas hydrate. According to G.Borman et al. (E.F.Shnyukov et al., 2015).

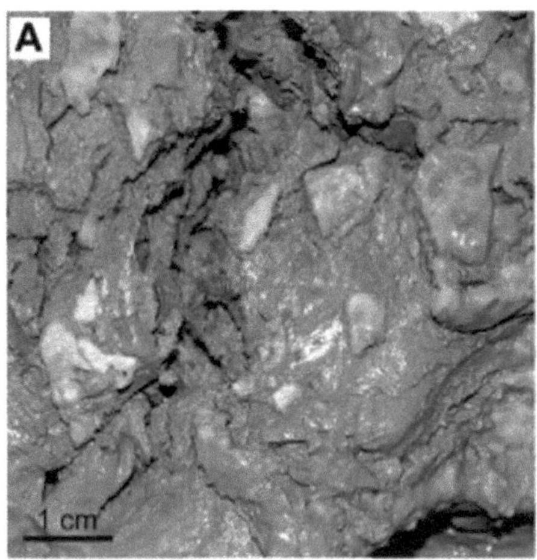

Fig. 17. Gas hydrate inclusions in the sopochnaya breccia of the Vodyanitsky mud volcano. Based on the materials of the R/V Meteor cruise, 2007. (E.F. Shnyukov et al., 2015).

It is interesting to compare the composition of gases in gas hydrates from different points of the Black Sea. In our opinion, the chemistry of gas hydrates, more precisely the gas composition, is very important, because, firstly, it indicates to a certain extent the thermodynamic conditions of formation and thus narrows or expands the areas of potential areas of gas hydrate development and their position in the sediment section, and, secondly, in the presence of CO_2, can, say, determine the occurrence of carbonate structures above the gas outflow structures. The latter can to some extent be search criteria for the detection of gas outflow points and gas hydrates. The composition of gases has not been studied in all cases of gas hydrate finds, but still a certain data bank has been accumulated. These data are contradictory in many respects. Perhaps, this indicates the diversity of gas composition of gas hydrates at different points. According to the gas composition, Yu.A.Byakov and R.P.Kruglyakova (2001) distinguish two types of gas hydrates. At the mud volcanoes MSU, Strakhov, Bezymyannyi, referred to the first type, methane in gas hydrates is contained in the amount of 93.3-95.7%, its homologs - in the amount from 4.3 to 6.7%. In other analyses from the same points nitrogen (0.7-1.8%), carbon dioxide (0.85%), hydrogen sulfide (0.25%) are found. The second type - gas hydrates of the Vassoevich mud volcano - contains only methane.

23

According to G.D.Ginsburg, A.N.Kremlev, M.N.Grigoriev (1990) gas hydrates from the sopochnaya breccia of a mud volcano contain 97-98% methane, 0.02-0.4% ethane, 0.5-0.9% CO_2 , 0.25% hydrogen sulfide.

As we can see, the composition of the Black Sea gas hydrates varies in rather wide ranges, but still usually the predominant part is methane with a small admixture of other gases. In several cases, the isotopic composition of gases was also studied. Thus, according to A.Yu.Byakov, R.P.Kruglyakova (2001), the carbon isotopic composition δ^{13} C - 61.80 to - 63.55‰, indicates the mixed (biochemical and thermocatalytic) nature of the hydrocarbon.

Methane gas hydrates occur at high pressures and low temperatures. This causes their development in permafrost zones or in the depths of seas and oceans.

The actual methane gas hydrates are gas-bearing ice. One cubic meter of ice contains 166 cubic meters of gas. When they write about gas hydrate deposits, they mean the rock - sand, sandstone, clay - saturated with methane gas hydrate in different degrees - 10%, 20%, 30%, 40-60%. The saturation varies widely. In the literature it is considered that rocks saturated with gas hydrates up to 60% create an impermeable cover, under which gases accumulate, creating subhydrate methane deposits. At the same time, veins, crystals, confluences of pure gas hydrates, etc., discussed above, may be isolated at localized points in the rock. Based on the columns of tubes lifted from the upper three meters of the caprock, we can only assume relatively low gas hydrate contents. Obviously, the more significant concentrations are localized deeper. It is necessary to determine what are the intervals of gas hydrate development, at what methane gas hydrate contents geophysical boundaries, such as, say, BSR, are broken off. It is very important to establish by experiments critical parameters of gas hydrate deposits in mud volcanoes.

According to V.P.Nomokonov and S.N.Stupak (1998), "the expected thickness of a gas hydrate deposit in the deep-water part of the (Black Sea) depression may reach 400 m for methane and 400-500 m for natural gas at a sea depth of 2100-2200 m". Obviously, in this case, the usual sedimentary strata section is meant.

O.D.Korsakov, Yu.A.Byakov, S.N.Stupak (1989) write that the depth of the hydrate formation zone along the geothermal gradient averages 400500 m below the bottom (maximum 800-1000 m). The formation of gas hydrates occurs mainly in Quaternary sediments (their thickness in the Black Sea depression is 1-3 km). Neogene sediments are the reservoir of gas hydrates only in the anticlinal arch and were observed by these authors in the Sorokina Trough and on the eastern slope of the Tuapse Trough. It is necessary to take into account the differences in the conditions of gas hydrates occurrence in deposits on continental slopes and deep-water areas of paleodolines and in deposits around mud volcanoes. Our findings are scattered over almost all geological structures of the Black Sea area: the West Black Sea depression, the Kerch-Taman and Sorokin troughs, the Tuapse and Giresun troughs and others. The explanation of this contradiction lies, obviously, in the confinement of gas hydrate finds to the sopochnaya breccia of mud volcanoes. This is a special type of gas hydrate outcrops. As early as 1988, G.D. Ginzburg et al. (1988) drew attention to the identification of a special submarine-mud-volcanic type of gas hydrate accumulations based on the data of the study of the Southern depression of the Caspian Sea. These accumulations are controlled by the crater field of mud volcanoes, they are shallow, because their roof is located practically at the bottom, they can have significant gas saturation. Subsequent researchers (G.D.Ginzburg et al., 1997) noted the most important role of mud volcanic waters in hydrate formation. Gas hydrates precipitate from filtered gas-saturated waters, formed from gas dissolved in water.

Mud volcanoes, given their deep emplacement (P.N. Kropotkin, B.M. Valyaev, 1980; A.M. Dmitrievsky, B.M. Valyaev, 2002, 2010, 2011, etc.), are most likely created by mantle plumes and deep feeding by hydrocarbon gases, which in the conditions of the deep sea leads to the creation of hydrocarbon deposits on different floors.) are created, most likely, due to mantle plumes and deep feeding with hydrocarbon gases, which in the conditions of the deep sea leads to the creation of hydrocarbon deposits on different sedimentary cover floors (under thermodynamically accessible conditions) and, in particular, near-surface deposits of methane gas hydrates. Sometimes they have the character of ring deposits, such as in the Haakon Mosby volcano in the Norwegian Sea

(V.A.Soloviev, 2001) and in the Black Sea Dvurechensky volcano (E.F.Shnyukov, A.P.Ziborov, 2004). At the same time, the distribution of gas outlets in and near the mud volcano is far from ideal and is often random. The occurrence of compensatory deflections near volcanoes causes numerous breaches of rock continuity, through which gases and water escape. It is the disturbance of the rocks of the tops of anticlines, on which mud volcanoes are developed, that can largely determine the distribution of gas hydrate deposits - the most gas hydrate-saturated rock sections.

As already mentioned, the occurrence of subsidence compensation deflections is a pattern of mud volcanic development in marine environments.

In the conditions of the Black Sea basin, inherited from Tethys, most of the volcanoes in the sea area can be assumed to have functioned episodically during at least the Neogene, up to the present day. In other words, the compensatory deflections should be filled with those Neogene and Quaternary sediments, at the times of accumulation of which the mud volcanoes were active. It is difficult to assume a dramatically different pattern of activity, given the observed effects in the Kerch-Taman region or in the present-day Black Sea in geologic history. Sea level fluctuations were large, but mainly affected the upper two hundred meters of the water column. The bulk of the presently observed volcanoes remained submerged. Sea depths varied considerably, but the water column remained. The emergence of compensatory deflections in case of volcanic activity was inevitable, as well as the accumulation of sediments in them. The predominance of clayey rocks alternating with hydrate-saturated rocks in the sediment section of the marine compensatory sags is very likely. The sediments will obviously reflect the peculiarities of sedimentation in different regions and zones of the paleomarine: in shallow-water areas - accumulation of coarser-grained, chemogenic or carbonate sediments, in the deep-water zone - of finer-grained sediments.

It is likely that the same compensatory deflections are likely to occur in the Caspian Sea, where large mud volcanoes are numerous, and in general in those areas of the World Ocean where mud volcanoes are found, for example, in the Mediterranean Sea, in the Atlantic near the island of Trinidad and elsewhere.

Compensatory deflections of deep-sea mud volcanoes can likely carry a payload. Mud volcanoes constantly emit natural gas mixtures in the form of flares somewhere in the center or in weakened zones near the volcanoes. Compensation sags are just such weakened zones. Therefore, at depths of more than 600-700 m, when pressures and temperatures favorable for the formation of gas hydrates were created, the most powerful accumulation of gas hydrates from the natural mixture of gases - methane, carbon dioxide, hydrogen sulfide - most likely took place in compensation structures. The concentration of gas hydrates is likely everywhere around the mud volcano, but it is in the compensation sags that the most significant accumulation is most likely to occur, these are some kind of traps.

When assessing potential gas hydrate accumulations, it is necessary to take into account the peculiarities of their development, established by V.A.Soloviev (2001): - the fact that only 1 to 10% of gases entering the hydrate formation zone are stabilized in gas hydrates;

- discreteness of gas hydrate development rather than the existence of their continuous covers;

- absence of widespread outlets and their presence only in some points;

- the fact that gas hydrates occupy only 10% of the total area of the hydrate formation zone;

- hydrate deposits are created by the flow of escaping gas-saturated water rather than by the flow of gases alone.

In this respect, the conclusion of A.I.Konyukhov, M.K.Ivanov, and L.M.Kulnitsky (1990) on the results of the study of the mud volcano Moscow University is not uninteresting:

"At the top of the Moscow University volcano, searches for gas hydrates were conducted back in 1988 immediately after its discovery, not only by MSU employees, but also by one of the marine geological parties of the Yuzhmorgeologiya association. All of them turned out to be fruitless, despite the fact that dozens of columns were raised and viewed. Apparently, gas hydrates here are confined to narrow zones where the actual sopochnaya breccia comes to the surface, i.e. where active mud volcanism continues. And since this process is narrowly localized, the zone of gas hydrates development is very limited in area and is probably associated with the activity of some peripheral crater in the western part of the volcano summit rather than the central one.

It should be assumed that such a narrow zone of methane gas hydrates development is the zone of compensatory deflection, to which the development of gas hydrates is confined. The thickness of sediments in the Novoselovskaya depression on the Kerch Peninsula reaches 211 meters. It is possible to assume the existence of not less thick clay-gas hydrate deposits in compensation troughs near large mud volcanoes of the Black Sea. Near the mud volcanoes in the section of modern Black Sea sediments the thickness of hydrotroilite silts sharply increases, which testifies to increased hydrogen sulfide intake from the subsurface. Hence the probability of occurrence of hydrogen sulfide gas hydrates. However, only methane hydrates are most probable. At the same time, one should not exclude clay particles adhering to gas bubbles and settling of even lighter hydrates CO_2 and $H_2 S$ to the bottom.

A detailed study of some acoustic profiles of the Black Sea bed allows us to fix small deflections-grabens on the sections of many mud volcanoes. A review of publications allows us to find such compensatory deflections or even grabens, for example, in the materials on the Black Sea and other authors. Compensation sags are a kind of traps for gas hydrates.

The development of subsidence phenomena can be observed at many volcanoes. Thus, a subsidence deflection is recorded on one of the Black Sea volcanoes. It is described by V.A.Gorchilin and L.M.Lebedev (1991) (Fig. 18). Such a sag was described by A.N.Gonchar et al. (2004) at the Tredmar mud volcano (Fig. 19). N.Kenyon, M.K.Ivanov, A.I.Akhmetzhanov et al. (E.F. Shnyukov et al., 2015) show a subsidence structure, near the NIOZ volcano (Fig. 20). A similar phenomenon was observed at the Neftyanoy volcano (Fig. 21), etc.

Of particular interest is a large ring structure in the central part of the West Black Sea depression, identified during the voyages of the R/V "Kiev" and "Professor Vodyanitsky". Its characterization in general terms is presented in Fig. 22-23. As can be seen, the Ring structure was crossed several times by transects of different directions and is currently drawn as a roughly oval basin framed by stepped ledges up to 30-70 m high in the section. The relief of the depressed central part of the basin is complicated by a series of small and low hills (up to 5-10). The diameter of the basin reaches 6.5 miles (about 12 km) along the elongation of the oval.

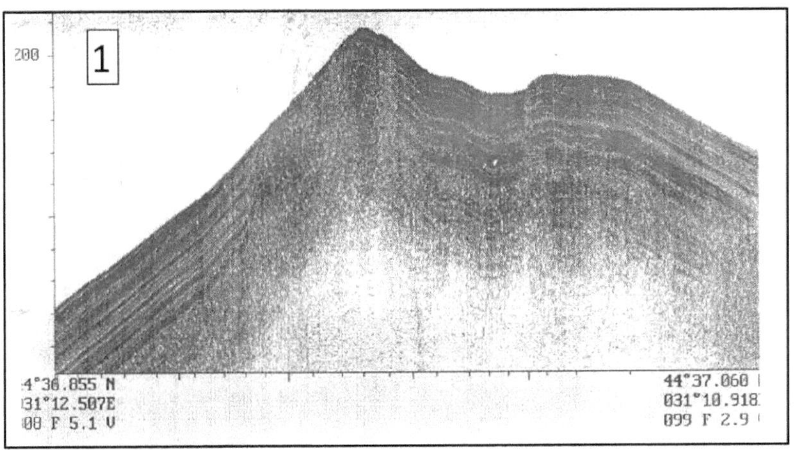

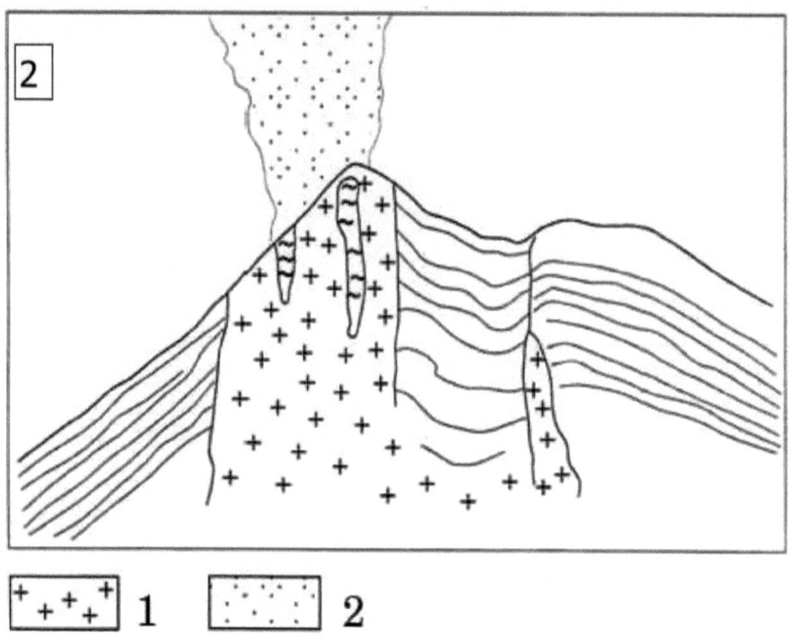

Figure 18: Acoustic profile through one of the Black Sea mud volcanoes.

1 - profile; 2 - profile interpretation

Notation: 1 - sopochnaya breccia; 2 - gas outlets.

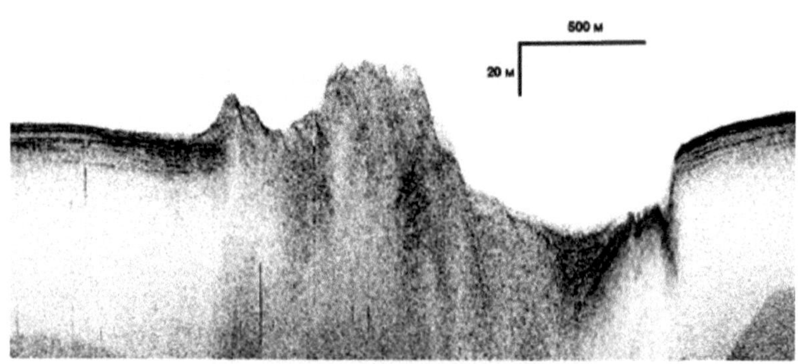

Fig. 19. Profilogram of Tredmar volcano. According to E.F. Shnyukov et al., 2015.

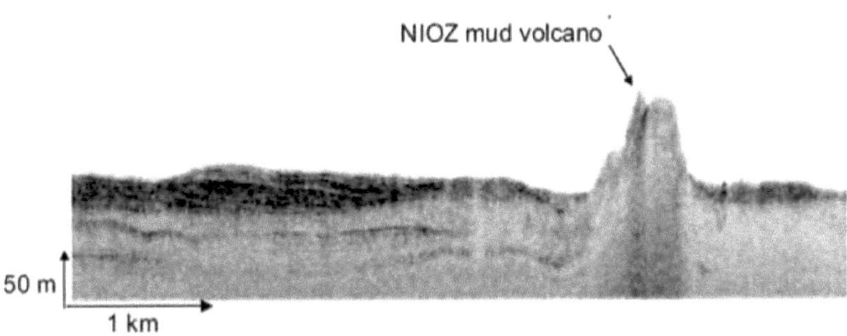

Fig. 20. Fragment of the bottom profile through the mud volcano **NIOZ**. By E.F. Shnyukov et al., 2015.

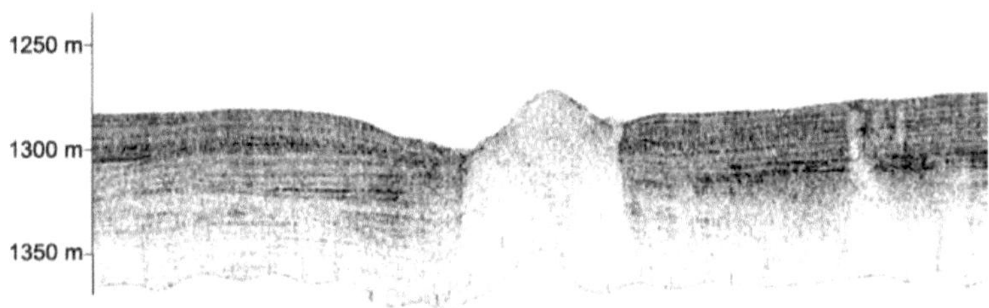

Fig. 21. Neftyanoy mud volcano. Sonogram and **acoustic** section. According to E.F. Shnyukov et al., 2015.

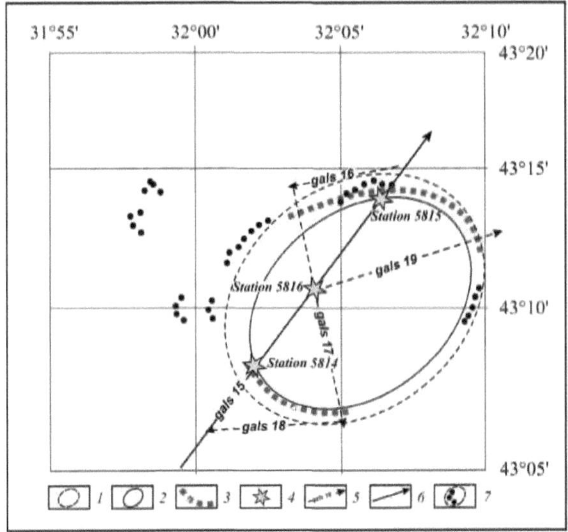

Figure 22: Scheme of research vessel transects through the Ring Structure. E.F. Shnyukov et al. 2015.

Notation: *1* - external boundaries of the morphostructure; *2* - boundaries of internal ledges; *3* - boundaries of the morphostructure, established; 4 - sampling stations; *5* - sampling routes and tack numbers; 6 - line of the profile of hydroacoustic sampling of GAK "Buk" in the voyage of the R/V "Kiev"; 7 - ledges identified during the 69th voyage of the R/V "Professor Vodyanitsky".

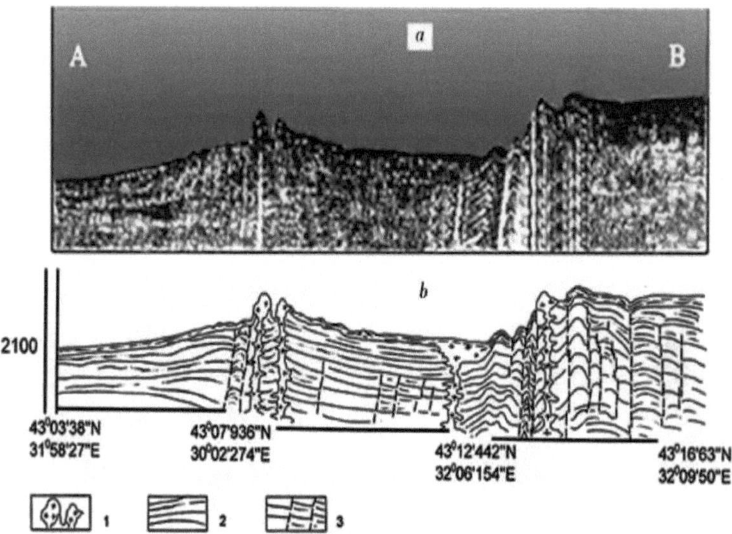

Fig. 23. Acoustic profile and profile interpretation through the Ring structure in the West Black Sea Basin. E.F. Shnyukov et al. 2015.

Several bottom tubes that allowed taking columns of bottom sediments up to 4 m long in the basin revealed an unusual section, in particular three cycles of accumulation of hydrotrolitic silts (0.97-1.45 m, 1.55-2.10 m, 2.22-3.0 m) overlain by sapropels and sapropelic silts (0.28-0.7 m) and coccolithic silts from above (Fig. 24). Typical sapropelic breccia was not encountered. Similar sections with increased thicknesses of hydrotroilitic silts were found earlier at some large mud volcanoes.

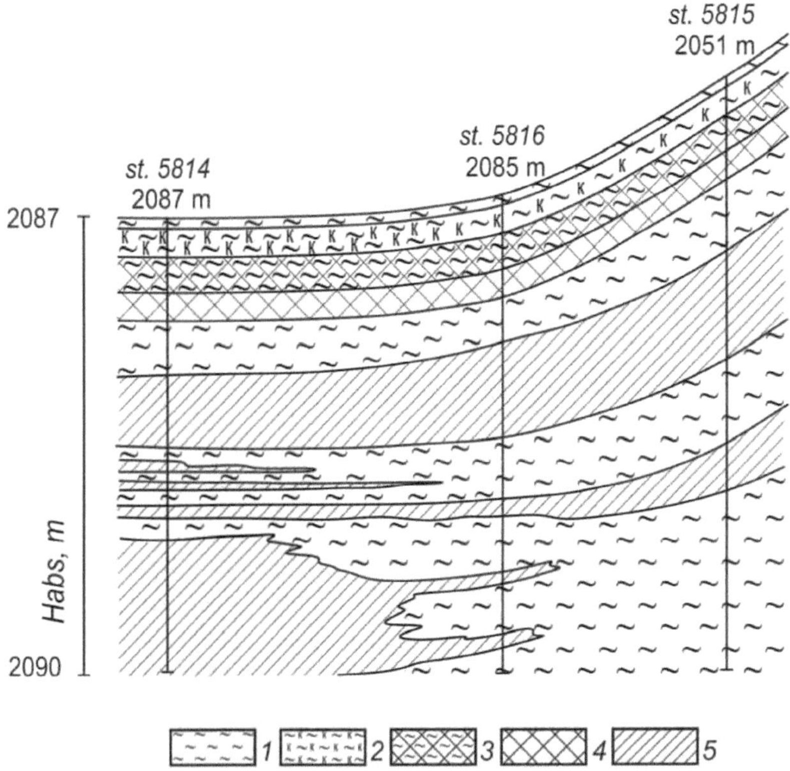

Fig. 24. Lithologic section through the morphostructure. E.F. Shnyukov et al. 2015.

Notation: 1 - silt; *2* - coccolithic silt; *3* - sapropelic silt; *4* - sapropel; *5* - hydrotroilite silt.

Different explanations of the nature of the ring structure are possible. According to a number of indirect signs, it is possible to assume the mud volcanic origin of the identified morphostructure. This is evidenced by the relief, specific layering and dislocation of sediments observed in the section obtained by the hydroacoustic complex "Buk" on the R/V "Kiev", some lithological features, high gas saturation of silts. Intensive degassing of the uplifted sediments indicates that the bottom of the basin is a so-called "gas swamp", which is typical for mud volcanic fields and gas saturated muds of the Black Sea.

A possible geologic interpretation of the acoustic profile, suggests the presence of several mud volcanoes on top of a large very gentle positive structure.

The revealed morphostructure by its geomorphologic and structural features resembles the Bulganak mud volcanic center on the Kerch Peninsula. The marginal ramparts and scarps bordering both basins and the shallow-flattened relief of the basin created by mud volcanoes are clearly comparable. It is very likely that the basin was formed as a result of powerful and long-lasting mud volcanic processes. This probably explains the sagging character of the central part of the rounded basin, which is comparable to a huge subsidence structure (depressed syncline).

Judging by the presence of a number of conical hills on the bottom near the basin, this is a whole new area of mud volcanoes.

Taking into account the already mentioned connection of gas hydrates with the activity of mud volcanoes, the presence of constant gas release in the basin, we can assume the existence of an extensive gas hydrate deposit within the studied morphostructure.

The greatest interest when working in the Black Sea is most likely not in the modern activity of mud volcanoes, but rather in rather ancient centers functioning, say, during the Neogene and up to the present day. In this case, false tectonic compensation structures can reach significant sizes and accumulate sediments containing gas hydrates of considerable thickness.

Since, however, gas outcrops around mud volcanoes are not necessarily localized only in troughs - compensatory synclines, but are often developed over a larger area, the formation of gas hydrates will not be limited to these troughs either. In compensatory troughs it is probable that gas hydrate accumulations are simply more powerful and, possibly, more ancient.

However, gas hydrates react quite clearly to changes in the sea temperature regime and, quite probably, may not be preserved in the sections of Neogene sediments. This issue is subject to further study.

Thus, the study of mud volcanism in the Black Sea region suggests that:

-      that mud volcanoes are a search sign of gas hydrates, as it is shown in many publications;

- in mud volcanoes in the conditions of the deep Black Sea, peculiar special type deposits of methane gas hydrates are formed;

- following G.D. Ginzburg et al. (1992), it can be assumed that gas hydrate deposits of mud-volcanic type are developed in many other areas of the World Ocean;

- Most likely, by analogy with terrestrial mud volcanoes, the Maikop age of the Black Sea mud volcanoes can be assumed;

It is likely that near the Black Sea mud volcanoes there are false tectonic compensatory troughs similar to the Kerch "depressed synclines";

- it is possible to assume localization of gas hydrate deposits of greater thickness in these false tectonic sags than around mud volcanoes;

- The large Ring morphostructure in the center of the West Black Sea depression, where gas hydrate deposits are assumed to exist, is likely to be attributed to subsidence structures;

- development of similar subsidence synclines is likely at many active mud volcanoes and other areas of the World Ocean.

Conclusion

Mud volcanoes of the Black Sea are deep formations according to modern geophysical materials and the data of studies of fluidogenic mineralization of the pillar breccia. The activity of the Black Sea mud volcanoes, similar to the data on mud volcanoes in Azerbaijan, Kerch-Taman and other regions, is accompanied by the release of huge masses of sap gases, sap breccia, and clastic material, which creates a kind of mass deficit in the interior of the volcanoes and leads to subsidence phenomena in the form of gradients. In terrestrial conditions, erosion processes seem to mask and reduce the scale of subsidence phenomena, but in marine conditions these grabens are filled with marine sediments, which create additional pressure for their subsidence. A kind of false compensatory synclines emerge.

Mud volcanoes functioning in the Neogene on the Kerch Peninsula accumulated significant thicknesses of Neogene sediments in these false synclines. The coincidence of the territories of volcanic activity and iron ore sedimentation in the Cimmerian time (Pliocene stage) led to the accumulation of oolitic iron ores. There are 9 such deposits and occurrences on the Kerch Peninsula, including Novoselovskoye (150 million tons of ore), Baksinskoye (40 million tons of ore) and others. Mud-volcanic structures were the receptacle of sedimentary iron ores with estimated total reserves of 400-450 million tons. These phenomena are characteristic of the shallow Cimmerian Sea. In the present-day Black Sea, most of the volcanoes are confined to the deep-water part of the sea. Mud volcanoes lying at depths greater than 600 m are located in a zone thermodynamically favorable for methane hydrate formation. In 60% of volcanoes, crystals and accumulations of gas hydrates are found in the upper layer as part of the sopochnaya breccia. Mud volcanoes are actually a search criterion for the formation of methane gas hydrate deposits. The main gas hydrate deposits in the Black Sea are localized on continental slopes, in deep-water paleodeltas.

Mud volcanoes develop their own type of gas hydrate deposits confined to the volcanic vents; they were first recorded by G.D. Ginzburg et al. (1992) in the Caspian Sea.

In the Norwegian volcano Haakon Mosby the gas hydrate deposit forms a giant "bagel". We can expect ring forms of deposits in the Black Sea volcanoes as well. At the same time, the most significant accumulation of gas hydrates is likely to occur in subsidence grabens due to many rock disturbances. This is probably a kind of gas hydrate trap, where layers of gas hydrates (gas hydrate-saturated sediments or sap breccia) alternate with ore-free layers. Such phenomena are likely to occur not only in the Black Sea, but also in many other areas of mud volcano development in the World Ocean.

List of references

1. *Aedusin P.P.* Mud volcanoes of the Crimean-Caucasian geological province. - M.: Izd. of the USSR Academy of Sciences, 1948. - 192 c.

2. *Aliee Ad.A., Guliee IS, Dsikpiea* F.G *and dp.* Atlas of mud volcanoes of the world. - Baku: Nafta Press, 2015. - 323 c.

3. *Alyae* S.E. New data on the tectonics of the Kerch Peninsula // Izv. of the USSR Academy of Sciences. Ser. geol. - 1947. - № 6. - C. 97-99.

4. *EoHdapuyK V. G.* Geology of Ukraine. - K.: Vid-Vo AN URSR, 1959. - 830 c.

5. *Byakoye Yu.A., Kruglyakoya R.P.* Gas hydrates of the Black Sea sedimentary strata - hydrocarbon raw materials of the future // Exploration and protection of subsoil. - 2001. - № 8. - C. 14-19.

6. *Gaeriloye Y.V.* Principle of isostasy in the formation of mud volcanoes. - Baku: Izd. FAN. - 1939.

7. *Geological* Dictionary: In 2 vol. - M.: Nedra, 1973. - 456 c.

8. *Geological assessment of* the routes of communication lines Sevastopol-Evpatoria, Sevastopol-Kerch, Sevastopol-Poti. - Kiev 2003. - 178 c.

9. *Geology of the* Sevastopol-O.Zmeiny-Zatoka fiber-optic communication route zone. - Kiev, 2004. - 142 c.

10. *Ginzburg G. D., Gramberg* I. S., *Guliee* I. S. *and dp.* Submarine-mud-volcanic type of gas hydrate accumulations // DAN USSR. - 1988. - T. 300,№2.-C. 416-418.

11. *Ginzburg G.D., Huseynoye R.A., Dadashee A.A. and dp.* Gas hydrates of the Southern Caspian Sea / Izv. RAS. - Ser. geol. 1992. - № 7. - C. 5-20.

12. *Ginzburg G.D., Kremleye A.N., Grigor'ee M.N. and dp.* Filtrogenic gas hydrates in the Black Sea // Geology and Geophysics. - 1990. - №3. - C. 10-20.

13. *Goloyekinsky N.A.* Report of the hydrogeologist of the Tavricheskaya zemstvo for 1889. - Simferopol, 1890.- 54 p.

14. *Gorchilin V.A., Lebedee L.I.* On the signs of gas hydrates in the sedimentary strata of the Black Sea and a possible type of hydrocarbon traps // Geol.zhurn. -1991.-#5.-S.7 5-81.

15. *Gubkin I.M.* Oil in the USSR // Energy Resources of the USSR. -M.; L.: Izd. of the USSR Academy of Sciences, 1937.-T. 1.-C. 169-194.

16. *Dmitrievskiy A.N., Valyaev* B.M. Fluidogeodynamic and genetic aspects of hydrate-bearing section of the sediments of the bottom of the World Ocean // Geodynamics and oil-and-gas bearing structures of the Black Sea-Caspian region.-2002.-P. 58-59.

17. *Dmitrievsky* A.N. Formation of energy-active and fluid-saturated zones of the Earth / Degassing of the Earth: geotectonics, geodynamics, geofluids, oil and gas, hydrocarbons and life. - M.: Geos, 2010. - C. 5-6.

18. *Ivanov M.K., Konyukhov A.I., Kulchitsky* L.M. *et al.* Mud volcanoes in the deep-water part of the Black Sea // Vesn. Mosk. un-ta. Ser. geol. 1989. -№ 3. -C.48-54.

19. *Ismagilov A.F., Kozlov V.N., Terekhov A.A., Khortov A.V.* Clay diapirism and mud volcanism in the formation of local structures in the Russian part of the Black Sea // Geology, geophysics and development of oil and gas deposits. - 2006. - 2. - C. 4-10.

20. *Kalinko* M.K. Basic regularities of oil and gas distribution in the Earth's crust. - M.: Nedra, 1964. - 207 c.

21. *Kovalevskiy S.A.* Gas volcanism (volcanoes and volcanoids) // Azerb. neft. khozyoz-vO. -№ 1. - 1935. -40 c.

22. *Konyukhov A.M., Ivanov M.K., Kulnitsky A.M.* About mud volcanoes and gas hydrates in the deep-water areas of the Black Sea // Lithology and Mineral Resources. - 1990. - №3. - C. 12-23.

23. *Korsakov O.D., Byakov Yu.A., Stupak* S.N. Gas hydrates of the Black Sea depression // Sov. geologiya. - 1989. -№ 12. - C. 4-10.

24. *Kropotkin P.N., Valyaev* B.M. Geodynamics of mud volcanic activity (in

connection with non-tegaziferousness) // Geological and geochemical bases of oil and gas prospecting. - K.: Nauk. dumka, 1980. - C. 148-178.

25. *Lychagin G.A.* Fossil mud volcanoes of the Kerch Peninsula // BMOIP. - Dept. Geol. - 1952. - 27. - Vyp. 4. - C. 3-13.

26. *Maimin* Z.L. Tertiary sediments of the Crimea. - M.-L.: Gostoptekhizdat, 1951-230 p.

27. *Meisner* L.B., *Tugolesov D.A.* Fluidogenic deformations in the sedimentary fulfillment of the Black Sea depression // Exploration and protection of subsoil. - 1997. - № 7. - C. 18-21.

28. *Naumenko* P./7. Modern activity of mud volcanism of the Kerch Peninsula // Proceedings on mineralogy, petrography and geochemistry of sedimentary rocks and ores. - K.: Nauk. dumka, 1976. - Vol. 4. - C. 115-135.

29. *Nomokomov* S.//. *Stupak* S.N. Signs of gas hydrate deposits in the Black Sea // Izv. of universities. - Geol. and Exploration. - 1998. -№ 3. - C. 72-78.

30. *Plotnikov A. M.* Hydrocarbon losses on mud volcanoes of the Kerch Peninsula // Geology and oil and gas content of the Black Sea depression. - K.: Naukova Dumka, 1967. - C. 72-81.

31. *Prokopov K.A.* Geotectonic sketch of the Kerch Peninsula and its relation to the Crimea and Taman // Tr. Gl. geol.-exploration. administration. - 1931. - Issue. 38. -C. 13-23.

32. *Rakhmanov* R.R. Mud volcanoes and their significance in predicting gas and oil bearing capacity of the subsurface. - Moscow: Nedra, 1987. - 174 c.

33. *Solov'ev V.A.* Gas-hydrate bearing capacity of the World Ocean subsoil // Gas Industry. - 2001. - № 12. - C. 19-23.

34. *Stadnitskaya A.N., Belenkaya I.Yu.* Composition and origin of hydrocarbon gases and their influence on diagenetic carbonate formations (Sorokina Trough, NE part of the Black Sea) // Geology of the Black and Azov Seas. - K.: Gnosis, 2000. - C. 155-

163.

35. *Fedorov S.F.* Mud volcanoes of the Crimean-Caucasian geological province and diapirism / Results of the study of mud volcanoes of the Crimean-Caucasian geological province. - M.-L.: Izd. of the Academy of Sciences of the USSR, 1939.-P. 5-44.

36. *Kholodov* V.N. Mud volcanoes: regularities of location and genesis. Message 1. Mud volcanic provinces and morphology of mud volcanoes // Lithology and Mineral Resources. - 2002. - № 3. - C. 227-241. Message 2. Geological and geochemical features and formation model // Lithology and Mineral Resources. - 2002. - № 4. - C. 339-358.

37. *Shnyukov* E.F., *Ziborov A.P.* Mineral Riches of the Black Sea. - K.: "Carbon-Ltd", 2004. - 280 c.

38. *Shnyukov* E.F., *Kobolev V.P., Pasynkov A.A.* Gas volcanism of the Black Sea. - K.: Logos, 2013. - 384 c.

39. *Shnyukov E.F., Naumenko P.I.* Cimmerian iron ores of depressed synclines of the Kerch Peninsula. - Simferopol: Krymizdat, 1964.- 126 pp.

40. *Shnyukov E.F., Naumenko P.I., Lebedev Y.S. et al.* Mud volcanism and ore formation. - K.: Nauk. dumka, 1971. - 332 c.

41. *Shnyukov E.F., Stupina L.V., Rybak* E.N. *et al.* Mud volcanoes of the Black Sea. Catalog. - K.: Logos, 2015. - 254 c.

42. *Shnyukov E.F., Sobolevsky Y.V., Gnatenko G.I. et al.* Mud volcanoes of the Kerch-Taman region. Atlas. - K.: Nauk. dumka, 1986. - 148 c.

43. *Shnyukov E.F., Sheremetev V.M., Maslakov N.A. et al.* Mud volcanoes of the Kerch-Taman region. - Krasnodar: Glavmedia, 2005. - 176 c.

44. *Shteber E.A.* Mud volcano Karabetova Gora near Taman // Izv. of the Caucasian branch of the Russian Geographical Society. - 1909/1910. - T. 20.

45. *Yakubov A.A., Grigoryants B.V., Aliev A.D., et al.* Mud volcanism of the Soviet Union and its connection with oil and gas bearing capacity. - Baku: ELM, 1980. -165

c.

46. ΔW7/7∏α/7/7 *G.*, A⅛αraw *E.*, /C/r *A. g/ al.* Origin and distribution of methane and methane hydrates in the Black Sea. - Cruise No. 84, leg 2. - Istanbul. - 2011. - 59 p.

47. /.7/<⅛7α/7 *T.*, Ifawg *H.K.*, Ka/7m///7g *P. g/* al., 2004. Heat flow and quantity of methane deduced from a gas field in the vicinity of the Dnepr Canyon, northwestern Black Sea // Geo-Mar. Lett. - 2004. - 24. - P. 182-193.

48. Poj9β5cw I., Zer/colα/5 G., Paw/w *N. g/* al. Late Quaternary channel avulsions on the Danube deep-sea fan. Black Sea // Mar. Geol. - 179 (12). - 2001. - P. 25-37.

Evgeny Fedorovich Shnyukov

Black Sea mud volcanoes as a search sign of methane gas hydrates

In this paper, modern ideas about mud volcanoes and accompanying subsidence structures are presented. In shallow sea conditions in the geological past, iron ores accumulate in mud volcanic structures, while in the conditions of the modern Black Sea in its deep water part, methane gas hydrates can accumulate in mud volcanoes, and especially actively in the accompanying subsidence structures. This is a special type of gas hydrate accumulation.

The work is intended for a wide range of specialists interested in the problem of methane gas hydrates.

Key words: mud volcanoes, gas emissions, subsidence structures, gas hydrates

I want morebooks!

Buy your books fast and straightforward online - at one of world's fastest growing online book stores! Environmentally sound due to Print-on-Demand technologies.

Buy your books online at
www.morebooks.shop

Kaufen Sie Ihre Bücher schnell und unkompliziert online – auf einer der am schnellsten wachsenden Buchhandelsplattformen weltweit! Dank Print-On-Demand umwelt- und ressourcenschonend produzi ert.

Bücher schneller online kaufen
www.morebooks.shop

Printed by Books on Demand GmbH, Norderstedt / Germany